YOUR KNOWLEDGE HAS VALUE

AF131378

Anuj Dutt

Arduino based RADAR System

GRIN Verlag

Bibliografische Information der Deutschen Nationalbibliothek:

Die Deutsche Bibliothek verzeichnet diese Publikation in der Deutschen National-
bibliografie; detaillierte bibliografische Daten sind im Internet über http://dnb.d-
nb.de/ abrufbar.

Imprint:

Copyright © 2014 GRIN Verlag GmbH
Druck und Bindung: Books on Demand GmbH, Norderstedt Germany
ISBN: 978-3-656-68039-0

This book at GRIN:

http://www.grin.com/en/e-book/275114/arduino-based-radar-system

Arduino Based RADAR System

By Anuj Dutt

Abstract— RADAR is an object detection system which uses radio waves to determine the range, altitude, direction, or speed of objects. The radar dish or antenna transmits pulses of radio waves or microwaves which bounce off any object in their path. Arduino is a single-board microcontroller to make using electronics in multidisciplinary projects more accessible. This project aims at making a RADAR that is efficient, cheaper and reflects all the possible techniques that a radar consists of.

I. INTRODUCTION

RADAR is an object detection system which uses radio waves to determine the range, altitude, direction, or speed of objects. Radar systems come in a variety of sizes and have different performance specifications. Some radar systems are used for air-traffic control at airports and others are used for long range surveillance and early-warning systems. A radar system is the heart of a missile guidance system. Small portable radar systems that can be maintained and operated by one person are available as well as systems that occupy several large rooms.

Radar was secretly developed by several nations before and during the World War II. The term RADAR itself, not the actual development, was coined in 1940 by United States Navy as an acronym for Radio Detection and Ranging.

The modern uses of radar are highly diverse, including air traffic control, radar, astronomy, air-defense systems, antimissile systems, antimissile systems; marine radars to locate landmarks and other ships; aircraft anti-collision systems; ocean surveillance systems, outer space surveillance and rendezvous systems; meteorological precipitation monitoring; altimetry and flight control precipitation monitoring; altimetry and flight control systems; guided missile target locating systems; and ground-penetrating radar for geological observations. High tech radar systems are associated with digital signal processing and are capable of extracting useful information from very high noise levels.

II. LITERATURE SURVEY

A. "The Idea"

Army, Navy and the Air Force make use of this technology. The use of such technology has been seen recently in the self parking car systems launched by AUDI, FORD etc. And even the upcoming driverless cars by Google like Prius and Lexus.

This setup can be used in any systems the customer may want to use like in a car, a bicycle or anything else. The use of Arduino in this provides even more flexibility of usage of the above-said module according to the requirements.

The idea of making an RADAR came as a part of a study carried out on the working and mechanism of "Automobiles of Future". Hence this time I was able to get a hold of one of the Arduino boards, Arduino UNO R3. So, knowing about the power and vast processing capabilities of the Arduino, I thought of making it big and a day to day application specific module that can be used and configured easily at any place and by anyone.

Moreover, in this fast moving world there is an immense need for the tools that can be used for the betterment of the mankind rather than devastating their lives.

Hence, from the idea of the self driving cars came the idea of self parking cars. The main problem of the people in the world is safety while driving. So, this gave up a solution to that by making use of this project to continuously scan the area for traffic, population etc. and as well as protection of the vehicles at the same time to prevent accidents or minor scratches to the vehicles.

III. COMPONENTS REQUIRED

A. Arduino UNO R3 or Above

The Arduino Uno is a microcontroller board based on the ATmega328. It has 14 digital Input /Output pins (of which 6 can be used as PWM outputs), 6 analog inputs, a 16MHz ceramic resonator, USB connection, a power jack, an ICSP header and a reset button. It contains everything needed to support the microcontroller; simply connect it to computer with a USB cable or power it with a AC-to-DC adapter or battery to get started.

The Uno differs from all preceding boards in that it does not use the FTDI USB-to-serial driver chip. Instead, it features the Atmega16U2 programmed as a USB-to-serial converter.

Changes in Uno R3:
1. Pin out: added SDA and SCL pins that are near to the AREF pin and two other new pins placed near to the reset pin, the IOREF that allow the shields to adapt to the voltage provided from the board. In future, shields will be compatible with both the board that uses the AVR, which operates with 5v and with the Arduino due that operates with 3.3v.
2. Stronger RESET circuit.
3. ATmega16U2 replace the 8U2.

"Uno" means one in Italian and is named to mark the upcoming release of Arduino 1.0. The Uno and version 1.0 will be the reference versions of Arduino, moving forward. The Uno is the latest in a series of USB Arduino boards, and the reference model for the Arduino platform; for a comparison with previous versions, see the index of Arduino Boards.

Figure 1.2 ATmega328P

C. Crystal Oscillator

A crystal oscillator is an electronic oscillator circuit that uses the mechanical resonance of a vibrating crystal of piezoelectric material to create an electrical signal with a very precise frequency. This frequency is commonly used to keep track of time (as in quartz wristwatches), to provide a stable clock signal for digital integrated circuits, and to stabilize frequencies for radio transmitters and receivers. The most common type of piezoelectric resonator used is the quartz crystal, so oscillator circuits incorporating them became known as crystal oscillators, but other piezoelectric materials including polycrystalline ceramics are used in similar circuits.

Quartz crystals are manufactured for frequencies from a few tens of kilohertz to hundreds of megahertz. More than two billion crystals are manufactured annually. Most are used for consumer devices such as wristwatches, clocks, radios, computers, and cell phones.

Quartz crystals are also found inside test and measurement equipment, such as counters, signal generators, and oscilloscopes.

Figure 1.1 Arduino Board

TABLE I
ARDUINO UNO R3 SPECIFICATIONS

Microcontroller	ATmega328
[1] Operating Voltage	5V
[2] Input Voltage (recommended)	7-12V
[3] Input Voltage (limits)	6-20V
[4] Digital I/O Pins	14 (of which 6 provide PWM output)
[5] Analog Input Pins	6
[6] DC Current per I/O Pin	40 mA
[7] DC Current for 3.3V Pin	50 mA
[8] Flash Memory	32 KB (ATmega328) of which 0.5 KB used by bootloader
[9] SRAM	2 KB (ATmega328)
[10] EEPROM	1 KB (ATmega328)
[11] Clock Speed	16 MHz

Figure 1.3 Crystal Oscillator(16 MHz)

D. Servo Motor

A servomotor is a rotary actuator that allows for precise control of angular position, velocity and acceleration. It consists of a suitable motor coupled to a sensor for position feedback. It also requires a relatively sophisticated controller, often a dedicated module designed specifically for use with servomotors.

Servomotors are not a different class of motor, on the basis of fundamental operating principle, but uses servomechanism to achieve closed loop control with a generic open loop motor.

Servomotors are used in applications such as robotics, CNC machinery or automated manufacturing.

B. ATmega328P

The ATmega328 is a single chip micro-controller created by Atmel and belongs to the mega AVR series. The high-performance Atmel 8-bit AVR RISC-based microcontroller combines 32 KB ISP flash memory with read-while-write capabilities, 1 KB EEPROM, 2 KB SRAM, 23 general purpose I/O lines, 32 general purpose working registers, three flexible timer/counters with compare modes, internal and external interrupts, serial programmable usart, a byte-oriented 2-wire serial interface, spi serial-port, a 6-channel 10 bit Analog to Digital converter (8-channels)in tqfp and qfn/mlf packages),programmable watchdog timer with internal oscillator and five software selectable power saving modes. The device operates between 1.8-5.5 volts. By executing powerful instructions in a single clock cycle, the device achieves throughputs approaching 1 MIPS per MHz, balancing power consumption and processing speed.

Figure 1.4 Servo Motor

E. Voltage Regulator

A voltage regulator is an electrical regulator designed to automatically maintain a constant voltage level.

With the exception of shunt regulators, all modern electronic voltage regulators operate by comparing the actual output voltage to some internal fixed reference voltage. Any difference is amplified and used to control the regulation element. This forms a negative feedback servo control loop. If the output voltage is too low, the regulation element is commanded to produce a higher voltage

The 78XX series of three-terminal positive regulator are available in the TO-220/D-PAK package and with several fixed output voltages, making them useful in a wide range of applications. Each type employs internal current limiting, thermal shut down and safe operating area protection, making it essentially indestructible. If adequate heat sinking is provided, they can deliver over 1A output current.

Although designed primarily as fixed voltage regulators, these devices can be used with external components to obtain adjustable voltages and currents.

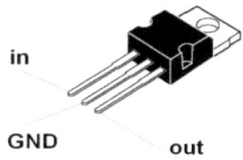

Figure 1.5 Voltage Regulator

F. Ultrasonic Sensor

Ultrasonic sensors (also known as transceivers when they both send and receive, but more generally called transducers) work on a principle similar to radar or sonar which evaluate attributes of a target by interpreting the echoes from radio or sound waves respectively. Ultrasonic sensors generate high frequency sound waves and evaluate the echo which is received back by the sensor. Sensors calculate the time interval between sending the signal and receiving the echo to determine the distance to an object.

This technology can be used for measuring wind speed and direction (anemometer), tank or channel level, and speed through air or water. For measuring speed or direction a device uses multiple detectors and calculates the speed from the relative distances to particulates in the air or water.

To measure tank or channel level, the sensor measures the distance to the surface of the fluid. Further applications include: humidifiers, sonar, medical ultra sonography, burglar alarms and non-destructive testing.

Systems typically use a transducer which generates sound waves in the ultrasonic range, above 18,000 hertz, by turning electrical energy into sound, then upon receiving the echo turn the sound waves into electrical energy which can be measured and displayed.

Figure 1.6 Ultrasonic Sensor

IV. USING ARDUINO SOFTWARE

The Arduino integrated development environment (IDE) is a cross-platform application written in Java, and is derived from the IDE for the Processing programming language and the Wiring projects. It is designed to introduce programming to artists and other newcomers unfamiliar with software development. It includes a code editor with features such as syntax highlighting, brace matching, and automatic indentation, and is also capable of compiling and uploading programs to the board with a single click. A program or code written for Arduino is called a "sketch".

Arduino programs are written in C or C++. The Arduino IDE comes with a software library called "Wiring" from the original Wiring project, which makes many common input/output operations much easier.

Users only need define two functions to make a run able cyclic executive program:

- Setup(): a function run once at the start of a program that can initialize settings
- Loop(): a function called repeatedly until the board powers off.

Open the Arduino IDE software and select the board in use. To select the board:

- Go to Tools.
- Select Board.

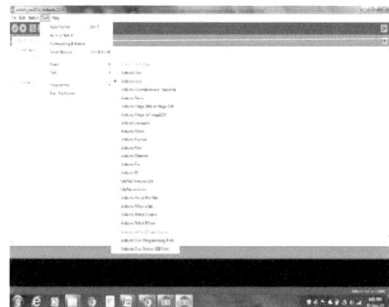

Figure 1.7 Arduino IDE

- Under board, select the board being used, in this case Arduino Uno.
- Go to Tools and to Port and select the port at which the Arduino board is connected.
- Write the code in the space provided and click on compile. Once the code is compiled, click on upload to upload the sketch to the Arduino board.

B. Connecting Servo Motor

V. PRACTICAL IMPLEMENTATION

A. Making On Arduino Board/ Boot-loading ATmega328P

Since, we believe in learning by doing. So, we decided to make our own arduino board instead of using the readymade board. So, the steps required to make an arduino board are as follows:

> Boot-loading an Atmega328 using the Arduino board/AVR Programmer by uploading the boot loader to the Microcontroller.

Figure 1.8 Boot-loading ATmega328P

> Making the connections on a general purpose PCB, connecting the crystal osicillator, capacitors, connectors for the connections to Arduino board etc.
> Providing the power supply, usually 5 volts.
> Arduino is Ready to use.

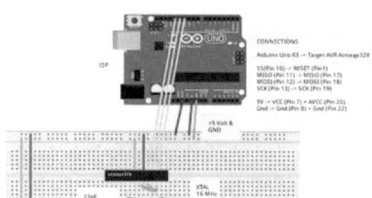

Figure 1.9 Boot-Loader Circuit

After you have done all this, then only the minimum circuitry like crystal oscillator, capacitors, connectors, power supply is required to complete the board. The same circuit can be made on the PCB, either designed or general purpose. Since, Arduino is an Open-Source. Hence, it is easy to make and can have any enhancements as per the requirements.

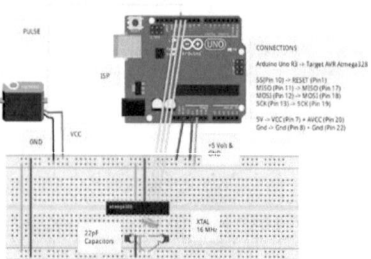

Figure 1.10 Connecting Servo Motor

A servomotor is a rotary actuator that allows for precise control of angular position, velocity and acceleration.

A normal servo motor has three terminals:
1. VCC
2. GND
3. PULSE

A servo motor works at normally 4.8 to 6 volts. Ground is provided by connecting it to the Ground of the Arduino. The total time for a servo motor pulse is usually 20ms. To move it to one end of say 0 degree angle, a 1ms pulse is used and to move it to other end i.e 180 degrees, a 2ms pulse is applied. Hence, according to this to move the axis of the servo motor to the center, a pulse of time 1.5 ms should be applied. For this, the pulse wire of the servo motor is connected to the Arduino that provides the digital pulses for pulse width modulation of the pulse. Hence, by programming for a particular pulse interval the servo motor can be controlled easily.

C. Connecting Ultrasonic Sensor

An Ultrasonic Sensor consists of three wires. One for Vcc, second for Ground and the third for pulse signal. The ultrasonic sensor is mounted on the servo motor and both of them further connected to the Arduino board. The ultrasonic sensor uses the reflection principle for its working. When connected to the Arduino, the Arduino provides the pulse signal to the ultrasonic sensor which then sends the ultrasonic wave in forward direction. Hence, whenever there is any obstacle detected or present in front, it reflects the waves which are received by the ultrasonic sensor.

If detected, the signal is sent to the Arduino and hence to the PC/laptop to the processing software that shows the presence of the obstacle on the rotating RADAR screen with distance and the angle at which it has been detected.

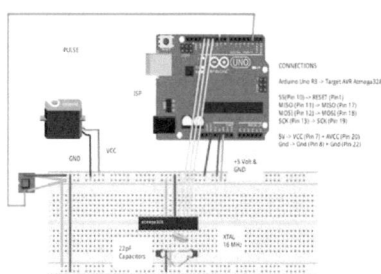

Figure 1.10 Connecting Ultrasonic Sensor to Arduino

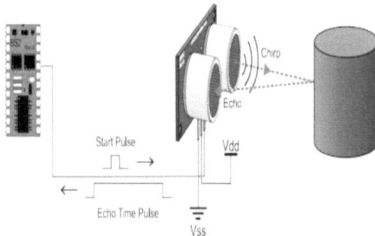

Figure 1.11 Working of Ultrasonic Sensor

VI. USING PROCESSING SOFTWARE

Processing is an open source programming language and integrated development environment (IDE) built for the electronic arts, new media art, and visual design communities with the purpose of teaching the fundamentals of computer programming in a visual context, and to serve as the foundation for electronic sketchbooks. The project was initiated in 2001 by Casey Reas and Benjamin Fry, both formerly of the Aesthetics and Computation Group at the MIT Media Lab. One of the stated aims of Processing is to act as a tool to get non-programmers started with programming, through the instant gratification of visual feedback. The language builds on the Java language, but uses a simplified syntax and graphics programming models.

Figure 1.12 Processing 2.0 Software

VII. PROBLEMS FACED

A. Making Own Arduino Board

The Arduino boards are available readily in the electronics market, but we decided to make our own Arduino board instead of buying one. So, the first problem was where to start from to achieve this goal. Since, all parts on an Arduino board are SMD's, so we had to find a way to replace the SMD's with DIP IC's and also had to make an AVR programmer in order to pursue our further work. Hence, it took us some days to determine and plan our course of action.

After that we had to boot load the AVR chip so as to make it compatible with the Arduino IDE software. Hence, we had to find a way to boot load the Arduino using the AVR programmer. It took us a long time to make the AVR programmer by researching on the type of communication and architecture of the AVR as it is not as same as a 8051 microcontroller.

B. Communicating with Arduino through PC

Another major problem related to the Arduino board was the communication with it from PC. Since, there is a requirement of an RS-232 to TTL conversion for the communication, so try some methods:

[1] Firstly I used the MAX-232 IC to communicate with the Arduino as with the 8051 but due to large voltage drop and mismatch in the speed, it failed to communicate.

[2] Next, I tried to use a dedicated AVR as USB to Serial converter as in the original Arduino board, the difference being DIP AVR used by us instead of the SMD Mega16U2 controller. But, unfortunately I was unable to communicate through it.

[3] At last I had no other choice but to use the FTDI FT-232R chip for USB to Serial conversion. Finally IT WORKED!!!

VIII. APPLICATIONS

The idea of making an Ultrasonic RADAR appeared to us while viewing the technology used in defense, be it Army, Navy or Air Force and now even used in the automobiles employing features like automatic/driverless parking systems, accident prevention during driving etc. The applications of such have been seen recently in the self parking car systems launched by AUDI, FORD etc. And even the upcoming driverless cars by Google like Prius and Lexus.

Figure 1.13 Driverless Car by Google

A. Air Force

In aviation, aircraft are equipped with radar devices that warn of aircraft or other obstacles in or approaching their path, display weather information, and give accurate altitude readings. The first commercial device fitted to aircraft was a 1938 Bell Lab unit on some United Air Lines aircraft. Such aircraft can land in fog at airports equipped with radar-assisted ground-controlled approach systems in which the plane's flight is observed on radar screens while operators radio landing directions to the pilot.

Figure 1.14 RADAR in Air Force

B. Naval Applications

Marine radars are used to measure the bearing and distance of ships to prevent collision with other ships, to navigate, and to fix their position at sea when within range of shore or other fixed references such as islands, buoys, and lightships. In port or in harbor, vessel traffic service radar systems are used to monitor and regulate ship movements in busy waters.

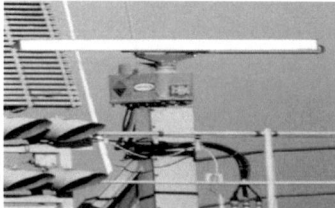

Figure 1.15 RADAR in Navy

C. Applications in Army

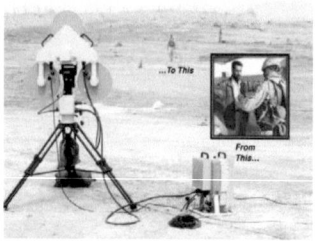

Figure 1.16 RADAR in Army

Two video cameras automatically detect and track individuals walking anywhere near the system, within the range of a soccer field. Low-level radar beams are aimed at them and then reflected back to a computer, which analyzes the signals in a series of algorithms. It does this by comparing the radar return signal (which emits less than a cell phone) to an extensive library of "normal responses."

Those responses are modeled after people of all different shapes and sizes (SET got around to adding females in 2009). It then compares the signal to another set of "anomalous responses" – any anomaly, and horns go off.

Literally, when the computer detects a threat, it shows a red symbol and sounds a horn. No threat and the symbol turns green, greeting the operators with a pleasant piano riff.

IX. A FINAL LOOK

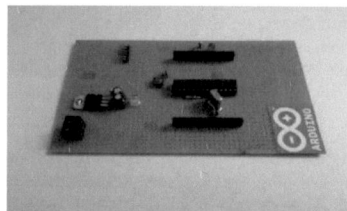

Figure 1.17 Arduino Board (Self Made)

Figure 1.17 Final Project

7

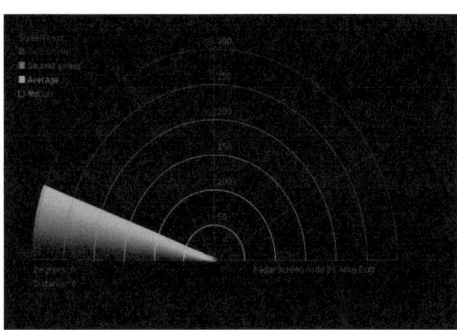

Figure 1.18 RADAR Screen using Processing 2.0

bombers-from-100-yards/
[13] http://upload.wikimedia.org/wikipedia/commons/Radara
ccumulationseng.png
[14] http://arduino.cc/en/Tutorial/BarGraph/
[15] http://arduino.cc/en/Tutorial/LiquidCrystal/
[16] http://fritzing.org

X. CONCLUSION

This project aims on the use of Ultrasonic Sensor by connected to the Arduino UNO R3 board and the signal from the sensor further provided to the screen formed on the laptop to measure the presence of any obstacle in front of the sensor as well as determine the range and angle at which the obstacle is detected by the sensor. For this screen, we use Processing 2 software by Ben Fry and Casey Rease, Massachusetts Institute of Technology, Cambridge.

Also, in addition to it, a set of LED's connected through the shift register and a buzzer also tells about the range of the obstacle. According to the range of the object, the green, yellow and red LED's glow up with variations in the buzzer output.

REFERENCES

[1] http://www.arduino.cc/
[2] http://www.arduinoproducts .cc/
[3] http://www.atmel.com/atmega328/
[4] http://en.wikipedia.org/wiki/File:16MHZ_Crystal.jpg
[5] http://www.google.co.in/imgres?imgurl=http://www.elec
trosome.com/wp-
content/uploads/2012/06/ServoMotor.gif&imgrefurl=http
://www.electrosome.com/tag/servomotor/&h=405&w=45
8&sz=67&tbnid=rcdlwDVt_x0DdM:&tbnh=100&tbnw=
113&zoom=1&usg=__6J2h0ZocdoSMrS1qgK1I2qpTQS
I=&docid=lEfbDrEzDBfzbM&sa=X&ei=a_OKUvTbD8
O5rgeYv4DoDQ&ved=0CDwQ9QE
[6] http//:www.sproboticworks.com/ic%20pin%20configurat
ion/7805/Pinout.jpg/
[7] http://www.sproboticworks.com/ic%ultrasonicsensor%2
0pinout.jpg
[8] http://www.instructables.com/id/ ATMega328-using-
Arduino-/
[9] http://www.motherjones.com/files/blog_google_driverles
s_car.jpg
[10] http://www.google.co.in/imgres/Radar_antenna.jpg&w=
546&h=697&ei=wuuK
[11] http://www.radomes.org/museum/photos/equip/ANSPS1
7.jpg
[12] http://www.wired.com/dangerroom/2011/07/ suicide-